Interlinking of Rivers in India: A Socio-economic Environmental evaluation

Dr. Hemant Pathak

DEDICATION

Dedicated to Shri Sainath Maharaj the all omnipotent of world the most merciful.

CONTENTS

Foreword

The National River Linking Project (NRLP) was proposed as the solution to water-related problems in India. It envisages transferring the waters of the water surplus basins to the water deficient basins in the south and the west. Today India coping with annual floods and droughts, both occurring at the same time in different regions, has been a major concern for India.

These concerns are more acute today as the growing population and the resultant increase in water demand place a heavy burden on the unevenly distributed water resources, and also cause huge economic losses to the financially vulnerable groups of the population.

This ambitious project initiated at that time when there was strong opposition to large dams. Environmentalists questioned on the ecological cost of large dams, while the NGOs and civil society probed the social cost of people displacement.

This book provides an essential guide to researchers, it offers: various aspects of National River Linking Project in present scenario.

Simply explained, Interlinking of Rivers in India: A Socio-economic Environmental evaluation is an important book bringing together diverse viewpoints from opponents and state agencies and regulators, for all who wish to make a difference in how to plan, conserve and manage our Rivers water.

Dr. Hemant Pathak

M.Sc. (Gold medalist), Ph. D.

Assistant Professor of Engineering Chemistry

Indira Gandhi Govt. Engineering college, Sagar, MP, India

Acronyms

EA   Environmental Assessment

EIS   Environmental Impact Statement

EPA   Environmental Protection Agency

IWMI   International Water Management Institute

NRLP National River Linking Project

NWDA National Water Development Agency

Glossary

Act A law

Bed The bed (also called the river bed) is the bottom of the river

Climate change A regional change in temperature and weather patterns. Current science indicates a discernible link between climate change over the last century and human activity, specifically the burning of fossil fuels.

Cost-benefit analysis A technique designed to determine the feasibility of a project or plan by quantifying its costs and benefits

Ecosystem An interactive system that includes the organisms of a natural community association together with their abiotic physical, chemical, and geochemical environment.

Environmental impact assessment (EIA) An environmental impact assessment (EIA) is an analytical process or procedure that systematically examines the possible environmental consequences of the implementation of a given activity (project). The aim is to ensure that the environmental implications of decisions related to a given activity are taken into account before the decisions are made.

Environmental policy A policy initiative aimed at addressing environmental problems and challenges.

Estuary Area at the mouth of a river where it broadens into the sea, and where fresh and seawater intermingle to produce brackish water. The estuarine environment is very rich in wildlife, particularly aquatic, but it is very vulnerable to damage as a result of human activities.

Eutrophication The degradation of water quality due to enrichment by nutrients, primarily nitrogen and phosphorus, which results in excessive plant (principally algae) growth and decay. Eutrophication of a lake normally contributes to its slow evolution into a bog or marsh and ultimately to dry land. Eutrophication may be accelerated by human activities that speed up the ageing process.

Forest

Land spanning more than 0.5 hectares with trees higher than 5 metres and a canopy cover of more than 10 per cent, or trees able to reach these thresholds in situ. It does not include land that is predominantly under agricultural or urban land use.

Forest degradation

Changes within the forest that negatively affect the structure or function of the stand or site, and thereby lower the capacity to supply products and/or services.

Global warming

increase in the average temperature of the earth's surface.

Groundwater

Water that flows or seeps downward and saturates soil or rock, supplying springs and wells. The upper surface of the saturate zone is called the water table.

Habitat

(1) The place or type of site where an organism or population naturally occurs. (2) Terrestrial or aquatic areas distinguished by geographic, abiotic and biotic features, whether entirely natural or semi-natural.

Policy

Any form of intervention or societal response. This includes statements of intent, such as a water policy or forest policy, Policy can be seen as a tool for the exercise of governance. When such an intervention is enforced by the state, it is called public policy.

Poverty

The pronounced deprivation of well-being.

River

A river is a large, flowing body of water that usually

empties into a sea or ocean.

Source   The source is the beginning of a stream or river.

Tributary   A tributary is a river or stream that flows into another stream, river, or lake.

Vulnerability   An intrinsic feature of people at risk. It is a function of exposure, sensitivity to impacts of the specific unit exposed (such as a watershed, island, household, village, city or country), and the ability or inability to cope or adapt. It is multi-dimensional, multidisciplinary, multisectoral and dynamic. The exposure is to hazards such as drought, conflict or fluctuations, and also to underlying socio-economic, institutional and environmental conditions.

Water quality   The chemical, physical and biological characteristics of water, usually in respect to its suitability for a particular purpose.

Water scarcity   Occurs when annual water supplies drop below 1 000 m3 per person, or when more than 40 per cent of available water is used.

Water stress   Occurs when low water supplies limit food production and economic development, and affect human health. An area is experiencing water stress when annual water supplies drop below 1 700 m3 per person.

1. Introduction

It is paradoxical to see floods in one part of the country while some other parts face drought. This drought-flood phenomenon is a recurring feature. Increasing demand for water at global, regional, national and local levels, Water is a scarce commodity and several basins such as Cauvery, Yamuna, Sutlej, Ravi and other smaller inter-State/intra-State rivers are short of water. 99 districts of the country are classified as drought prone; an area of about 40 million hectare is prone to recurring floods.

Rapidly increasing population of India and the resultant increase in water demand place a heavy burden on the unevenly distributed water resources, consequently huge economic losses to the financially vulnerable groups of the population. A huge demand to enhance and diversify food production to meet the needs of a vast population with changing consumption patterns and higher disposable incomes.

Designed to address these concerns, the National River Linking Project (NRLP) envisages transferring water from the potentially water surplus Himalayan rivers to the water-scarce river basins of western and peninsular India.

The Indian Rivers Inter-link is a proposed large-scale civil engineering project that aims to link Indian rivers by a network of reservoirs and canals and so reduce persistent floods in some parts and water shortages in other parts of India and free the country from the endless cycle of flood and drought.

Storage reservoirs on these rivers and connect them to other parts of the country, regional imbalances could be reduced significantly and lot of benefits by way of irrigation, domestic and industrial water supply, hydropower generation, navigational facilities etc. would accrue.

The dream of interlinking rivers is now being realised, with the Godavari formally connected to the Krishna (the second and the fourth longest rivers in the country) linked through a canal in Andhra Pradesh. It is aimed broadly at harnessing part of Godavari floodwaters which otherwise flow into the Bay of Bengal. This linkage is expected to create an irrigation potential of about 2.8 lakh hectares, besides supplying water for domestic and industrial uses in Krishna and West Godavari districts. The project was completed at a cost of Rs1,300 crore.

A second scheme, the Ken-Betwa river project estimated to cost Rs11,676 crore is currently under development. The total cost of the project includes three components: 1) the peninsular component will cost Rs.1, 06,000 crore, the Himalayan component will cost Rs.1, 85,000 crore, and the hydroelectric component will cost Rs. 2, 69,000 crore. The quantity of water diverted in the peninsular component will be 141 cubic kilometers and in the Himalayan component it will be 33 cubic kilometers.

Projects opponents are concerned about knowledge gap on environmental, ecological, social displacement impacts as well as unseen and unknown risks associated with playing with nature.

Some States like Rajasthan, Gujarat and Tamil Nadu have over expected from project and demanded reservation due to Water deficiency, ultimately supported the concept of inter-linking of rivers.

2. Overview

The average rainfall in India is about 4,000 billion cubic meters or about 1 million gallons of fresh water per person every year. Precipitation pattern in India varies dramatically across distance and over calendar months. These comes over a 4-month period June through September.

Rain pattern in the nation is not uniform, the east and north gets most of the rain, while the west and south get less. The northeastern region of the country receives heavy precipitation, in comparison with the northwestern, western and southern parts.

Spatial inequality too is extreme, Ganga-Brahmaputra-Meghna basins, which cover one third of the country's total land area, are home to 44 % of India's population, but drain more than 60 % of the country's water resources.

In contrast, the Krishna, Cauvery and, Penner river basins and the eastward flowing rivers between Penner and Kanyakumari cover 16 % of the total land area, host 17 % of the population, but drain only 6 % of India's water resources.

Owing to these unequal endowments, India's river basins are at different degrees of closure. Needs of these areas can be addressed, it is argued, by augmenting their natural flows through the transfer of surplus waters from the Himalayan Rivers.

Our nation evident of excess monsoons and floods from years, followed by below average or late monsoons with droughts. India is the second largest populations in world have year round demand for irrigation, drinking and industrial water creates a demand-supply gap has been worsening problems.

Diverting a portion of the surplus flood waters from the Himalayan rivers into the drought-prone areas can change the destiny of that areas. Annual floods, on average, affect more than 7 million ha of the total land area, 3 million ha of the cropped area and 34 million people, mostly in the eastern parts, and inflicts an annual damage of well over Rs.1,000 crores.

In contrast, recurrent droughts affect 19 % of the country, 68 % of the cropped area and 12 % of the population. The reservoir storages and the canal diversions in ILR are expected to reduce flood damages by 35 % and ease drought-proneness in semi-arid and arid parts, besides making 12 km^3 of water available for domestic and industrial water supplies in these drought-prone districts.

The need of the hour is to have a linking river, which will enable availability of water to the fields, villages, towns and industries throughout the year even while maintaining environmental purity.

Need of Interconnection of 37 Himalayan and peninsular rivers for surplus rivers will be dammed, and the flow will be diverted to rivers that could do with more water.

There would be requirement of networking of canals, small and large reservoirs will be constructed for hydroelectric power generation.

According to the NWDA, India will require about 450 million tonnes of food grain per annum to feed a population of 1.5 billion in the year 2050. For these requirements, there is a need to expand its irrigation potential to 160 million hectares, which is 20 million hectares more than the total irrigation potential.

3. History

The Inter-linking of Rivers in India proposal has a long history. While the timeline has the history since the project's conception 125 years ago. In British colonial rule at 19th century engineer Arthur Cotton proposed the plan to interlink Indian rivers in order to hasten import and export of goods from its colony in South Asia, as well as to address water shortages and droughts in southeastern India, now Andhra Pradesh and Orissa.

In 2002, the then President of India Abdul Kalam mentioned the river linking project during a speech. He proposed it as a solution to India's water woes after which an application requesting an order from the Supreme Court on that matter was submitted. The application was converted into a writ petition and finally, in October 2002, the Supreme Court ordered the Central Government to initiate work on inter-linking the major rivers of the country.

In the same year, a task force was appointed and a deadline of 2016 was set to complete the entire project that would link 37 rivers but nothing concrete happened until almost a decade ago. On Feb 27, 2012, the Supreme Court ordered the constitution of a "Special Committee for Interlinking of Rivers" headed by the Minister of Water Resources.

4. Scope of the Project

NWDA has studied and prepared reports on 14 inter-link projects for Himalayan component, 16 inter-link projects for Peninsular component and 37 intrastate river linking projects.

The National River Interlinking Project will comprise of 30 links to connect 37 rivers across the nation through a network of nearly 3000 storage dams to form a gigantic South Asian Water Grid. Inter-link project is conceptualized in two distinct components- the Himalayan and peninsular.

The former will transfer 33 Km^3 of water, and the latter will transfer 141 Km^3 of water through a combined network of 14,900 km long canals.

The Himalayan Component (HC), with 16 river links, has two sub-components, the first will transfer the surplus waters of the Ganga and Brahmaputra rivers to the Mahanadi Basin and from there the water will be relayed to Godavari, Godavari to Krishna, Krishna to Pennar and Pennar to the Cauvery basins.

The second sub-component will transfer water from the eastern Ganga tributaries to benefit the western parts of the Ganga and the Sabarmati river basins. Altogether, these transfers will mitigate floods in the eastern parts of the Ganga Basin, and provide the western parts of the basin with irrigation and water supplies.

The Himalayan component needs several large dams in Bhutan and Nepal to store and transfer flood waters from the tributaries of the Ganga and Brahmaputra rivers, and also within India to transfer the surplus waters of the Mahanadi and Godavari rivers.

Peninsular Rivers Development Component or the Southern Water Grid, which includes 16 links that propose to connect the rivers of South India. It envisages linking the Mahanadi and Godavari to feed the Krishna, Pennar, Cauvery, and Vaigai rivers.

This linkage will require several large dams and major canals to be constructed. Besides this, the Ken river will also be linked to the Betwa, Parbati, Kalisindh, and Chambal rivers.

The peninsular component has 16 major canals and four sub-components:

I. linking the Mahanadi-Godavari-Krishna-Cauvery-Vaigai rivers;

II. linking west flowing rivers that are south of Tapi and north of Bombay;

III. linking the Ken-Betwa and Parbati-Kalisindh-Chambal rivers; and

IV. diverting the flow in some of the west flowing rivers to the eastern side. The en route irrigation under the peninsular component is expected to irrigate a substantial area as proposed under the NRLP. This area to be irrigated is situated in arid and semi-arid western and peninsular India.

Table-1

Northern Himalayan Rivers inter-link component

Fourteen inter-links under consideration for Himalayan component (with feasibility study status identified) are as follows:

- Ghaghara–Yamuna link
- Sarda–Yamuna link
- Yamuna–Rajasthan link
- Rajasthan–Sabarmati link
- Kosi–Ghaghara link
- Kosi–Mechi link
- Manas–Sankosh–Tista–Ganga link
- Jogighopa–Tista–Farakka link
- Ganga–Damodar–Subernarekha link
- Subernarekha–Mahanadi link

- Farakka–Sunderbans link
- Gandak–Ganga link
- Chunar–Sone Barrage link
- Sone dam–Southern tributaries of Ganga link

The aim is to conserve monsoon flows for irrigation and hydropower generation, along with flood control. The linkage will transfer surplus flows of the Kosi, Gandak and Ghagra to the west. A link between the Ganga and Yamuna is also proposed to transfer the surplus water to drought-prone areas of Haryana, Rajasthan and Gujarat.

Inter-basin water transfer aims to transfer water from the surplus rivers to the deficit areas.

The project is being managed by India's National Water Development Agency (NWDA), under Ministry of Water Resources.

5. Potential Benefits of Indian rivers inter-linking projects

These projects may be solution of India's water Crisis, is to conserve the abundant monsoon water bounty, store it in reservoirs, and deliver this water to scarce areas.

Beyond water security, this linkage provide transport infrastructure through navigation, as well as to broadening income sources in rural areas through fish farming.

Following key points is important to considerable, expressed the importance of inter-linking projects.

- **Recurrence Floods and Drought**

The uncertainty of monsoons, sometimes marked by prolonged dry spells and fluctuations in seasonal and annual rainfall is a serious problem for the country.

Floods are a recurring feature, particularly in the Brahmaputra and the Ganga, affecting the States of Assam, Bihar, West Bengal and Uttar Pradesh, while large areas in the States of Rajasthan, Gujarat, Andhra Pradesh, Karnataka and Tamil Nadu that face recurring droughts.

India sees cycles of drought years and flood years, with large parts of west and south experiencing more deficits and large variations, resulting in immense hardship particularly the poorest farmers and rural populations. Lack of irrigation water regionally leads to crop failures and farmer suicides.

- **Shortage of drinking water**

India currently stores only 30 days of rainfall, while developed nations strategically store 900 days worth of water demand in arid areas river basins and reservoirs.

India's dam reservoirs store only 200 cubic meters per person at present and declining with increasing population. Despite abundant rains during July–September, some regions in other seasons see shortages of drinking water. Neglecting storage creation, resulting in economic water scarcity that may impede its economic growth.

Other countries of the world have invested heavily in storage creation; the U.S.A has a per capita storage capacity of 5,961 m^3 ; Australia has 4,717 m^3 , and Brazil has 3,388 m^3 .

China has increased its per capita storage capacity to 2,486 m^3. It is imperative that India increases its storage for regulating the vast amount of runoff that otherwise cannot be beneficially utilized.

- **Population and food security**

Population increase in India is the other driver of need for river inter-linking, which is around 1,200 million at present, is expected to increase to 1,500 to 1,800 million in the year 2050.

India's population growth rate has been falling, but still continues to increase by about 10 to 15 million people every year, that means the requirement of food grains will go up to about 450 million tonnes The resulting demand for food must be satisfied with higher yields and better crop security, both of which require adequate irrigation of about 140 million hectares of land.

It will be necessary to raise the irrigation potential in the country to 160 million hectares for all crops by 2050.Currently, just a fraction of that land is irrigated, and most irrigation relies on monsoon. River inter-linking is claimed to be a possible means of assured and better irrigation for more farmers, and thus better food security for a growing population.

- **Reduce regional imbalances**

Some years, the problem temporarily becomes too much rainfall, and weeks of havoc from floods. This excess-scarcity regional disparity and flood-drought cycles have created the need for water resources management.

The ambitious plan is to build storage reservoirs on these rivers and connect them to other parts of the country. Additionally, this will reduce regional imbalances.

Other benefits of the project spoken about are additional irrigation, domestic and industrial water supply, hydropower generation and navigational facilities.

- **Navigation**

India needs infrastructure for logistics and movement of freight. Using connected rivers as navigation is a cleaner, low carbon footprint form of transport infrastructure, particularly for ores and food grains.

- **Current reserves and loss in groundwater level**

India also relies excessively on groundwater, which accounts for over 50 percent of irrigated area with 20 million tube wells installed, satellite evidence of critical groundwater levels. About 15 percent of India's food is being produced using rapidly depleting groundwater.

The end of the era of massive expansion in groundwater use is going to demand greater reliance on surface water supply systems. India's water situation is already critical, and it needs sustainable development and management of surface water and groundwater usage.

- **Hydropower generation**

The river interlinking project claims to generate total power of 34,000 MW, 4 GW in the peninsular component and 30 GW in the Himalayan component. The addition of hydropower is expected to curb the drinking water woes of millions and supply water to industries in drought-prone and water-scarce cities in south.

- **Irrigation benefits**

The project claims to provide additional irrigation to 35 million hectares (m ha) in the water-scarce western and peninsular regions, which includes 25 m ha through surface irrigation and 10 m ha through groundwater.

This will further create employment, boost crop outputs and farm incomes and multiply benefits through backward by farm equipment and input supplies and forward linkages by agro-processing industries. Along with this the project is expected to create several benefits for navigation and fisheries.

Ken- Betwa Project proved to be useful for provide irrigation to the water scarce Bundelkhand region.

- **Livestock Population**

Livestock production, especially milk, is a major part of the agricultural economy in the Bundelkhand region will increased.

- **Laying of New Roads**

The project has the provisions for laying new roads in the project area for the easy communication and transportation of goods and materials needed for the various dam construction.

Once the roads are laid, these would be permanent in nature in most of the case. The laid roads, after the project also will create a road communication network in the area and would facilitate the locals and others for their smooth movements and also to start small scale industries.

- **Creation of Medical and Health Facilities**

The project proponents envisage the creation of medical and health facilities for the dam construction workers, staff, and employees during the implementation of project and project related activities. These medical and health facilities have the provisions of in-patient and out-patient treatment. After the project implementation also, the created medical and health facilities will not be withdrawn and could be utilized by the locals. This would bridge the gap of existing medical and health infrastructural facilities and to a largest extent would solve the problems of locals. Improvement in Sanitation Environment: With the implementation of EMP, takes care of sanitation and hygiene aspects for keeping the healthy environment in the project area. By this, wherever the project colonies are constructed for project staff and workers as well as office buildings are built, provisions are made for the safe disposal of solid waste and surface drainages. This will also motivate the locals to adopt hygienic and sanitary habits to keep the environment clean.

- **Ground Water Recharge**

The field survey in the study area revealed that ground water levels are deep during pre-monsoon period. Some of the groundwater sources mostly used for drinking water go dry during this period. Under the project a large water body will be coming up by constructing dam and will certainly recharge and increase the groundwater levels in the project area. This will help the farming community as well as other water users who depended on groundwater facility

The catchment area is spread over Panna, Chhatarpur, Damoh, Sagar, Satna, Katni, Narasingpur and Raisen districts of Madhya Pradesh. The dam site is in the Panna Tiger Reserve (PTR) and a part of the core area of PTR will get submerged due to the project.

The project has got good potential, particularly, because of close proximity of Daudhan dam site to Khajurao for recreation and tourism development. Provision for development of tourist huts, picnic spots has been made on the periphery of Rangwan reservoir, which is about 9 km from Daudhan dam site.

The link canal also offers good scope for tourism development. At the tail end of the canal, the Orcha temple (Jhansi) could become an ideal place for tourist. Therefore, it is proposed to develop tourist's huts and necessary provision has been made thereto.

- **Employment Generation**

This multipurpose Projects will generate employment at the time of construction and post construction phases. About 40 millian persons (including skilled and unskilled workers) will get employment under the project during the construction phase.

These projects will prove several economic benefits like development of agro-based industries, transportation and storage facilities. Increased in farm supplies, production and consumption of fertilizer, pesticide, farm equipment and employment generation.

Economic benefits of irrigation water supply include various benefits on, crop production; recharges groundwater; animal husbandry; farm equipments and agro-processing.

The activity will also generate employment during construction and operation phases of the project. This will benefit the economy, both national and local levels. Further, the increased agricultural production in the command area will stimulate the development of forward and backward linkages and in turn the economic development of the area.

The Panna Tiger Reserve area coming under submergence is mainly located in right flank of dam covered by steep hills, dense forest. Hence the area is not a major attraction for tourists. Presently most of the tourists visit the Pandava Falls and only Plateau regions of the National Park where open grass land with abundant numbers of herbivorous. These areas are in downstream of Daudhan dam and will not submerge. Reservoir is going to become a tourism attraction when impounded with water. There are several tourist spots around the proposed Daudhan area due to which the project site develops into a good tourist resort which is a positive impact.

Due to the implementation of Ken-Betwa Link Project Phase-I, it will give a good opportunity to the locals for getting employment during the construction period of the project. The locals would get preference for employment on daily wage basis as causal / workers.

6. Problems associated with Indian rivers inter-linking projects

India's ambitious plan to link its rivers has raised several ecological and social concerns, as well as questions about distribution. Now these fears are being echoed from across its borders.

Environmentalists pointed to the dangers of the seismic hazard, especially in the Himalayan component, and many worry about the transfer of river pollution that accompanies inter-basin water transfers and loss of forests and biodiversity. When a river flow outpours into the ocean, such flows are imperative to reduce salinity of oceans and to sustain estuaries and coastal habitats. The coastal and marine fisheries are affected, and rivers bring in filth.

Any human endeavour on such a scale is an artificial intervention in the ecology, ecological balance of land and oceans, fresh water and sea water, is also disrupted. But others argue differently. They opine that some Indian river basins have vast non-utilizable water resources, even after meeting all human and ecosystem services needs.

Globally, Bangladesh and Nepal have expressed concern over India's ambitious river-linking plan, fearing they may be denied their share of water flowing from the Himalayas.

Some scientists are worried about the ecological impacts of the project of such a massive scale. Government of India (May 2003) raised 23 environmental concerns about River linking.

Huge land required to construct diverting canals, problems of acquisition will be serious for government propose to acquire that land. Many irrigation projects have been stalled in India because of the detrimental effects on the environment that construction of large reservoirs and dams beget.

Disastrous impacts of blocking fresh water flows into the sea have reduced water outflows into the Sea, thus destroying deltaic mangroves. There is also a deep concern over linking polluted rivers to unpolluted rivers. Beyond the politics of water ownership and use, the matter of concern here is the condition of the dam itself. Kerala's rivers are part of unique, sensitive ecosystems, which would be disturbed by the river-linking project.

Gujarat had underlined the fact that water was a State subject and that the opinion of the respective States must be sought.

Bangladesh, in fact, so concerned about the repercussions of the move on their country that they are considering appealing to the United Nations to redraft international laws on water-sharing.

The Bangladesh government fears that India's plans to divert vast quantities of water from major rivers, including the Ganga and the Brahmaputra, will threaten the livelihoods of more than 100 million people downstream in Bangladesh.

Although India's plan to link major rivers flowing from the Himalayas and divert them south to drought-prone areas are still on the progress.

Bangladeshi scientists estimate that even a 10%-20% reduction in water flow to the country could dry out large areas for much of the year.

Linking of the Ken and Betwa rivers, at the Panna Tiger Reserve in Madhya Pradesh was expected to submerge an important wildlife habitat which was home to many endangered species. Apart from submersion, there were issues of noise pollution and movement of diesel vehicles which would kill the ecology of the forest.

The river has to flow; it has to reach the sea, even during the summer. The fishes that breed in the river and the people dependent on them need to be considered.

Many times a river development project by one state on an interstate river may submerge the territory of another riparian state. Hence, it becomes reasonable that the state that uses the territory of another state for storage purposes pays compensation to the latter.

Kerala, which has always been projected as a water surplus state, has linked rivers. The agreement between the two states, Kerala and Tamil Nadu, on the rights to compensation for land and water, sharing the benefits, power, and irrigation has been violated. The tribes which have been displaced from those areas were also not rehabilitated. our laws are not sufficient to address such conflict issues.

Another point of contention is the legal status of the water. In the Constitution of India, water is subject to state control, with the national government being allowed to intervene only in the regulation and development of interstate rivers to the extent that it is declared by the parliament to be a situation that is in public interest.

India is locked in conflicts both with its neighbours and domestically over water problems. Though the diplomatic relations of India with Nepal and Bangladesh are good, water issues could impinge upon these connections. Where the water comes from and where goes have become very critical issues in every water conflict.

Any withdrawal of water in the upper catchment area may cause depletion of water resources lower down the river. This may cause severe inter-district, inter-state, and inter-country disputes, as we witness in the cases of the Cauvery or Ganges.

Water is indeed in the State List when it comes to India, but this is subject to the provisions of Entry 56 in the Union List. The Sutlej–Yamuna link canal is an attempt at transferring the waters of River Sutlej to the Yamuna river basin.

In such cases, constructing a reservoir would add to the financial, social, and environmental costs. Though the government tries to justify the linking project, the impacts will be disastrous for states or countries living downstream as livelihoods are dependent on these rivers. For instance, the major economy of Bangladesh depends on the seasonal fluctuations of the Ganga and Brahmaputra. Also, wetlands such as the Sunderbans will be impacted by it.

Lalitpur, Madhya Pradesh with the largest dam density in Asia, is facing serious problems of depletion of ground water. While nearby Tikamgarh, there are 500-year-old ponds and small dams constructed by the Chola kings. These have retained the ground water level to assist 35% more land irrigation. The negative part of project is loss in crop and livestock production due to submergence of the crop area in the upstream of the reservoir. Malaria is already prevalent in the area. The condition during construction and operation phase will favour the spread of Malaria and gastroenteritis. There is scope for spread of STD and HIV diseases.

The proponents of the river linking project cite irrigation, more food production, and eventually, a solution to the future food security issues as major advantages. This needs to be looked at carefully.

Loss of Land

In all 6422.62 ha of private lands would be lost by the project affected families for the construction of Daudhan dam and other project components along with canal network under the project. Besides, as per the project design about 5339.00 ha of forest lands would also be brought under various project components. This indicates that the land owners would lose their landed properties.

Loss of Livelihood

About 72 per cent households would become landless, nearly 21 per cent would become marginal farmers and almost 7 per cent will fall under the small farmer category.

As a result of land acquisition the project affected big farmers would lose their big farmer status.

Loss of Employment

Due to land acquisition, several families, who become landless, would lose their total self employments, who otherwise have been engaged in their farming activity.

Loss of Income

Project affected household's socio-economic environment is going to affect their family life due to loss of land, livelihood and employment resulting in reduced regular family income.

Risks of transmigrations

Interlinking of rivers has been associated with various environmental issues. For instance, at the Suez Canal linking the Red Sea with the Mediterranean, some fish species of Red Sea origin passed through the canal and reached the eastern Mediterranean. The local species of fishes were displaced in this case. Such risks of transmigrations in interlinking of rivers are unknown.

7. Conclusion

Presented paper evaluated the socio-economic and other implications of the proposed cropping and irrigation patterns in the Ken-Betwa project. Water use for hydropower and in the domestic and industrial sectors were also not considered.

These would have generated significant benefits to the KBP region, as inadequate electricity and drinking water supply are major constraints for economic development in this region.

in severe drought years, some farmers sell their livestock as they are unable to provide an adequate drinking water supply for their livestock, let alone fodder and other feed.

Under Article 262, the Government of India created the Interstate Water Disputes Act of 1956 to resolve interstate problems, but water has become an increasingly politicized issue.

8. References

- International Water Management Institute, 2006, Strategic Analysis of India's National River Linking project (NRLP). http://nrlp.iwmi.org/PDocs/PrjProposal.asp.

- National Water Development Agency, 2006, www.nwda.gov.in.

- Alagh, Y.K., 2006, Methodology of Irrigation Planning. The Ken-Betwa Case. In Interlinking of Rivers in India. Overview and Ken-betwa Link, eds. respectively Yoginder K Alagh, Ganesh Pangare, and Biksham Gujja. New Delhi, India: Academic Foundation.

- Alagh, Y.K.; Pangare, G.; Gujja. B, 2006, Interlinking of Rivers in India. Overview and Ken-betwa Link. New Delhi, India, Academic Foundation.

- Bharatndu, P.; Shaailendra, N.G.; Santosh, S.; Phourasia, L.P., 1998, Problems and Potentials of Bundelkhand with Special Reference to Water Resource Base. Uttar Pradesh: V.S.K (Banda) publications.

- Chopra, K. 2006. The Feasibility Report of Ken-betwa Link Project: An Analysis of Assumptions and Methodology.

- In Interlinking of Rivers in India: Overview and Ken-betwa Link, eds.Yoginder K Alagh, Ganesh Pangare, and Biksham Gujja. New Delhi, India: Academic Foundation.

ABOUT THE AUTHOR

Dr. Hemant Pathak held positions as Assistant Professor in the department of hemistry, Govt. Indira Gandhi Engineering College, Sagar, MP, India. He had xtensive experience in teaching, research and administrative management.

Dr. Pathak received his Ph.D. degree in chemistry from Dr. Hari Singh Gour entral University, Sagar, India and M.Sc. Gold medalist from Jiwaji University, iwalior. He has published 23 books and more than 50 research papers in reputed nternational and National journals and received several awards. He is a member f editorial boards and reviewer boards of several international journals and ocieties. His area of specialization includes Engineering Chemistry, Energy udits and Environmental Pollution management.

www.ingramcontent.com/pod-product-compliance
Lightning Source LLC
Chambersburg PA
CBHW080529190526
45169CB00008B/3105